STARDUST SERENADE

The Cosmic Dance of the Perseid Meteor Shower

Mona B. Franklin

ISBN: 9798858040521

DEDICATION

To the Watchers of the Sky,

This book is a tribute to those who find beauty in the canvas of the
night sky, to the dreamers who trace constellations with their hearts,
and to the seekers who uncover the universe's secrets among the
stars. Your unwavering gaze uplifts us all and reminds us of the
infinite wonders that await above, including the ethereal magic of the
Perseid meteor shower.

With celestial appreciation,

Mona B. Franklin.

CONTENTS

1	Introduction	6
2	Origin of Perseid Meteor Shower	8
3	The Science behind Perseids	11
4	Viewing the Perseids	14
5	Notable Perseid Display in History	16
7	Perseids Popular Culture	18
8	Perseids and Space Exploration	21
9	Citizen Science and Perseids	23
10	Exploring the Future of Perseid Meteor Shower	25
11	How to Photograph the Perseids	27
12	Exploring Other Meteor Spectacles	29
13	Conclusion	31
14	Extras	33-35

INTRODUCTION

In the infinite expanse of the night sky, a spectacular performance unfolds, captivating the hearts and minds of stargazers and dreamers alike. This mesmerizing display, known as the Perseid meteor shower, transcends the boundaries of Earth and the sky, offering a glimpse into the grand cosmic choreography that has been enchanting humanity for centuries.

Picture a world bathed in the soft glow of moonlight, where stars twinkle like diamonds strewn across a velvet tapestry. Then, from the darkness emerges a streak of light, swift and radiant—a shooting star. This is the Perseid meteor, a celestial messenger that leaves an ephemeral trail of brilliance in its wake.

The Perseid meteor shower is nature's symphony, a harmonic convergence of Earth's orbit and the debris shed by Comet Swift-Tuttle, a cosmic traveller on a millennia-spanning journey. As our planet hurtles through space, it encounters the remnants of the comet—particles and dust that once danced around its nucleus. When these particles collide with Earth's atmosphere at mind-boggling speeds, they ignite in a fiery spectacle that graces our night skies.

Each August, the Perseids stage a captivating performance. The night transforms into a canvas, painted with streaks of light as meteor after meteor blazes a trail across the horizon. These meteors, though transient, leave an indelible mark on our memories, reminding us of the vastness of the universe and our place within it. From ancient civilizations to modern societies, the Perseid meteor shower has been a universal muse. In times long past, these celestial displays were interpreted as signs from the gods, omens that foretold the fate of nations and the destiny of individuals. The Perseids have inspired myths and legends, their radiance interwoven with the stories of heroes, lovers, and seekers of truth.

Gazing upward, we become part of an age-old tradition—a tradition of wonder, curiosity, and contemplation. The Perseid meteor shower beckons us to venture beyond our earthly concerns and connect with the cosmos.

But where and when can this spectacle be witnessed at its finest? How can we prepare ourselves to witness this cosmic dance in all its glory?

As we journey through this book, we will unravel the science, history, and magic of the Perseid meteor shower. We will explore its origins, delve into the art and literature it has inspired, and uncover the celestial mechanics that bring this awe-inspiring event to life. Whether you're an aspiring astronomer, an artist seeking inspiration, or simply a dreamer with eyes turned skyward, join us as we embark on a voyage to unravel the mysteries and beauty of the Perseid meteor shower.

Origins of the Perseid Meteor Shower

In the boundless cosmic theater, the Perseid meteor shower takes center stage as a celestial ballet that transcends time and space. To truly appreciate this breathtaking performance, one must venture behind the scenes and uncover the intricate choreography that has been captivating Earth's audience for eons. At the heart of the Perseid meteor shower lies a celestial wanderer known as Comet Swift-Tuttle.

This enigmatic traveler embarks on an elliptical journey that spans decades, bringing it close to the sun before sweeping out to the distant realms of the solar system. During its sojourns near our star, the comet releases a trail of dust and debris—tiny fragments that would otherwise remain hidden in the cosmic shadows. The remnants shed by Comet Swift-Tuttle, often referred to as meteoroids become celestial breadcrumbs that mark its passage.

Over centuries, these particles are scattered along the comet's path, forming a diffuse ribbon that spans millions of miles. When Earth's orbit intersects this celestial thoroughfare, the cosmic remnants plunge into our atmosphere, their high-speed collisions with air molecules creating the brilliant spectacle we know as meteors or "shooting stars."

The name "Perseid" pays homage to the constellation Perseus, the point in the sky from which these meteors appear to emanate. The radiant—the apparent origin of the meteor shower—adds a touch of mystique to the event. As Earth hurtles through space, the perspective from which we view the meteors change, creating the illusion that they are emerging from a single point in the heavens.

The name "Perseid" carries echoes of Greek mythology, where it finds its origins in the heroic tales of Perseus and his lineage. But this ethereal display is more than a mere tribute to ancient gods—it's a dance of light and fire that stretches across the heavens, captivating generations and uniting us with the cosmos. Derived from the Greek word "Perseidai," the Perseid meteor shower pays homage to the sons of Perseus, the mythical hero known for slaying the Gorgon Medusa.

Perseus' adventures intertwine with gods and monsters, their narratives etching their way into the constellations that adorn the night sky. The Perseids' radiant, a point from which meteors seem to originate, hovers near the constellations Perseus and Cassiopeia—a cosmic tribute to these timeless tales.

The Perseid meteor shower finds its celestial partner in Comet Swift-Tuttle, a cosmic traveler on a 133-year sojourn through our solar system. This comet, with its nucleus of ice and rock, leaves behind a trail of dust and particles during its close encounters with the sun. These particles, strewn across Swift-Tuttle's path, become the luminous meteors that grace our skies during the Perseid meteor shower.

The Perseid cloud, an ensemble of particles extending along Comet Swift-Tuttle's orbital path, showcases the cosmic legacy left by this celestial wanderer. Some of these particles have journeyed alongside the comet for a millennium, while others are more recent additions, ejected during Swift-Tuttle's passage in 1865. This mixture of ages creates a meteoric dance that unfolds across the ages, offering us glimpses into the comet's past and its intricate gravitational interactions with Earth.

Every August, our planet crosses paths with the Perseid cloud, unleashing a celestial spectacle that leaves us spellbound. As Earth plows through the cosmic debris, meteoroids collide with our atmosphere, creating luminous trails of light that streak across the night sky.

The Perseids' radiant presence bridges the gap between the terrestrial and the celestial, inviting us to look upward, to dream, and to connect with the wonders of the universe. The Perseids' radiant, a point from which meteors seem to originate, hovers of ice and rock, and leaves behind a trail of dust and particles during its close encounters with the sun.

These particles, strewn across Swift-Tuttle's path, become the luminous meteors that grace our skies during the Perseid meteor shower. When a Perseid meteor enters our atmosphere, it meets resistance from the air particles in its path. This friction generates intense heat, causing the meteoroid to incandescent and create the luminous streak we witness from the ground. These ethereal streaks, while fleeting, are a testament to the raw power of the cosmos, as well as a reminder of the intricate forces at play in the universe.

While the Perseid meteoroids are minuscule—often no larger than a grain of sand—the brilliance of their trails captivates our attention. This cosmic spectacle showcases the elegant dance between Earth and the remnants of a comet that has journeyed through space for millennia, leaving behind a luminous legacy that paints the night sky with fleeting strokes of brilliance.

The Science Behind Perseids

The cosmic symphony that is the Perseid meteor shower is composed of more than just ethereal beauty—it's a performance orchestrated by the intricate interplay between celestial bodies, the Earth's atmosphere, and the remnants of a distant wanderer.

At the heart of the Perseid meteor shower lies the tale of Comet Swift-Tuttle, a comet that embarks on a celestial odyssey that spans centuries. As Swift-Tuttle journeys through its elliptical orbit, it leaves behind a trail of dust and debris—a legacy that converges with Earth's orbit, giving birth to the radiant meteors that grace our skies. Understanding the trajectory of this celestial wanderer is key to unlocking the secrets of the Perseids.

The meteoroids that make up the Perseid meteor shower are not mere cosmic dust; they are remnants of Swift-Tuttle's icy nucleus. Composed of rock and ice, these particles represent frozen moments in time, encapsulating the formation of our solar system.

As they hurtle through space, they carry with them the story of their origin, and the scientific analysis of these meteoroids provides insights into the processes that have shaped our cosmic neighborhood.

As our planet traverses the Perseid meteor stream, a spectacular interaction unfolds at the boundary between Earth and space. These cosmic travelers, some as small as a grain of sand, collide with our atmosphere at extraordinary speeds. The resulting friction causes them to incandescent, creating the luminous streaks we witness as shooting stars. Understanding the dynamics of this interaction not only unveils the brilliance of the Perseids but also contributes to our comprehension of atmospheric processes and the mysteries of cosmic matter.

The study of meteor showers like the Perseids provides a unique window into the cosmos. By analyzing the composition, trajectory, and behavior of the meteoroids, astronomers gain insights into the nature of comets, the structure of our solar system, and the origin of cosmic matter. This understanding also sheds light on the broader universe, as meteoroids offer clues about the building blocks that form planets, asteroids, and other celestial bodies.

Through the lens of science, the Perseid meteor shower transforms from a poetic spectacle into a cosmic puzzle waiting to be solved.

The most striking instance of a meteor's influence on Earth is illustrated by the asteroid believed to be responsible for the renowned mass extinction of dinosaurs and numerous other species approximately 66 million years ago.

Occasionally, a meteor survives the descent and arrives as a substantial rock fragment known as a meteorite upon Earth's surface. This instantly becomes a valuable subject for researchers exploring the origins of the Solar System and life on our planet. Utilizing radioactive dating techniques, scientists can accurately estimate the age of meteorites, shedding light on the history of Solar System formation. The chemical compositions and ages of these meteorites, as well as Earth rock samples, provide support for the hypothesis that our planet formed from the same extraterrestrial matter that constitutes meteorites.

Meteorites have also supplied evidence of intricate organic compounds beyond Earth. By scrutinizing the chemical makeup of meteorites, scientists have identified molecules crucial to life, like amino acids. While this doesn't prove the existence of extraterrestrial life, it does indicate that the fundamental components for life can originate in space environments. Researchers studying organic compounds within meteorites even speculate that these rocks might have played a role in the inception of life on Earth.

So, as you observe the Perseids in the night sky, keep in mind that while they may appear as purely celestial occurrences, their ties to the Solar System's origins and life on Earth might make them more connected to our home planet than they initially seem.

Viewing the Perseids

The celestial theater of the Perseid meteor shower beckons observers to witness a cosmic spectacle that transcends time and space. From the darkness of the night sky, shooting stars streak across the heavens, leaving trails of brilliance that captivate our senses and spark our imagination.

As the Earth plunges through the Perseid meteoroid stream, the best times to view the meteor shower become a delicate dance between celestial mechanics and our own earthly perspectives. The peak of the Perseid meteor shower usually occurs between the nights of August 12 and 13, offering the prime opportunity to witness the most meteors. During these peak nights, up to 100 shooting stars an hour may grace the skies, creating a breathtaking show that leaves observers in awe.

To fully appreciate the Perseid meteor shower, finding the perfect vantage point is essential.

Seek out locations away from urban light pollution and tall structures that obstruct your view of the horizon. Dark skies, free from the interference of artificial lights, provide the ideal canvas for the meteoric spectacle to unfold. Whether on a hill, a remote field, or the shores of a tranquil lake, the key is to have an unobstructed view of the entire sky.

Unlike some astronomical events that require telescopes or specialized equipment, the Perseid meteor shower requires nothing more than your own eyes. Telescopes and binoculars can actually limit your field of view, making it more challenging to catch the shooting stars as they streak across the sky. The naked eye provides the widest perspective, allowing you to fully appreciate the luminous trails that illuminate the celestial canvas.

Observing the Perseid meteor shower is not only a visual treat but a sensory experience that engages all your senses. Bring along comfortable seating, blankets, and warm clothing to ensure your viewing session is both enjoyable and comfortable. Consider packing a thermos of hot drink to ward off the chill of the night, allowing you to focus solely on the cosmic ballet unfolding above. To fully immerse yourself in the Perseid meteor shower, consider these observing tips:

- Allow your eyes to adapt to the darkness by avoiding bright lights.

- Lie back and gaze upward, covering as much sky as possible.

- patience on your side—the meteors may appear in bursts of activity.

The Perseid meteor shower transforms the night sky into a canvas of light, a reminder of the cosmic dance that unites Earth and the universe. Armed with knowledge of the best times and locations for viewing, and armed with nothing more than your own eyes, you are poised to witness one of the most captivating astronomical events.

Notable Perseid Displays inHistory

Throughout the annals of time, the Perseid meteor shower has etched its radiant mark on the pages of human history. From ancient civilizations to modern societies, this cosmic phenomenon has left an indelible impression, shaping cultures, inspiring awe, and fueling the imagination.

As Earth crosses the path of the Perseid meteoroid stream year after year, it encounters a cosmic echo—a celestial symphony that has been playing for millennia. In the scrolls of antiquity, we find references to these brilliant meteors—ominous omens, messages from the gods, or signs of impending change. The Perseids' luminous trails have ignited the imaginations of ancient observers, weaving their brilliance into the fabric of myth and folklore.

In the tapestry of ancient cultures, the Perseids stand as celestial messengers, carrying potent symbolism and mythological significance. From the Greeks who named the meteor shower after the hero Perseus, to the Chinese who associated the Perseids with the tears of the Weaving Maid, the Perseid meteor shower has bridged the gap between the earthly and the divine. These cosmic displays were interpreted as the gods' communication with humanity, their messages etched across the canvas of the night sky.

As cultures evolved and civilizations flourished, the Perseid meteor shower continued to cast its celestial glow. In medieval Europe, the meteors were linked to the feast day of St. Lawrence, with legend suggesting they were the sparks of his martyrdom. The Perseids' luminous trails became a symbol of hope and inspiration, infusing stories, songs, and art with their brilliance. The annual cosmic show offered a shared experience that transcended geographical and temporal boundaries, uniting people in their wonder and curiosity.

With the advance of science and technology, the Perseid meteor shower transitioned from mysterious omens to understood phenomena. Observers no longer attributed these luminous trails solely to divine messages, but rather embraced them as cosmic gifts from the universe. The Perseids' annual return became a celebration of the beauty and grandeur of the cosmos, inspiring poets, writers, and artists to pay homage to their brilliance. The Perseid meteor shower stands as a testament to humanity's enduring connection to the cosmos.

As we navigate the chapters of history, we find ourselves gazing at the same celestial spectacle that our ancestors once beheld. The Perseids' radiance continues to inspire awe, curiosity, and a sense of shared wonder across cultures and generation.

Perseids Popular Culture

The Perseids constitute merely one instance of numerous bursts of shooting star occurrences throughout the year. Although astronomers only began to seriously study meteors in the 19th century, humans have observed this phenomenon for millennia.

Interestingly, the Perseids have been recognized by Catholics as the "tears of Saint Lawrence" for centuries. Legend has it that the saint, who was martyred by being burned alive on August 10 in 258 AD, sheds tears that remain suspended in the sky until descending to Earth annually on the anniversary of his martyrdom.

It's no surprise that before the era of modern science and astronomy, various cultures linked meteors to deities, often interpreting them as signs of impending divine anger or omens. In the realm of ancient astronomy, Roman historians noted "comet stars" in the August sky around 30 BC, seemingly related to the recent passing of Cleopatra; it's plausible that these observations were indeed ancient Perseids.

Similarly, observers in Japan, Korea, and China documented luminous phenomena and trails in the sky many centuries ago, with a Chinese account appearing to mention the Perseids around 36 AD.

Beyond the scientific intrigue and historical significance, the Perseid meteor shower has woven its radiant threads into the fabric of popular culture, leaving an indelible mark on literature, art, music, and modern entertainment.

Throughout literary history, the Perseid meteor shower has served as a metaphorical beacon, illuminating themes of wonder, transformation, and human connection. From ancient myths to modern novels, the shooting stars have symbolized everything from divine intervention to personal epiphanies. In works like Ray Bradbury's *"Falling Upward,"* the Perseids become a catalyst for introspection and discovery, a testament to the profound impact these cosmic wonders have on the human psyche.

The Perseids' luminous trails have not only graced the pages of books but have also found their way onto canvases and into sculptures. Artists have sought to capture the ethereal beauty of the meteor shower, creating visual representations that translate the awe and splendor of the cosmos onto a tangible medium. These artistic interpretations transcend time and space, inviting viewers to experience the brilliance of the Perseids through the lens of creativity.

From classical compositions to modern melodies, the Perseid meteor shower has inspired musicians to craft sonic tapestries that evoke the spirit of the cosmos. In music, the shooting stars become notes in a celestial score, creating harmonies that resonate with the universe's grandeur. Songs like Jonn Serrie's *"The Stars, Like Dust,"* pay homage to the Perseids, inviting listeners to close their eyes and journey among the stars.

The Perseid meteor showers' allure has not escaped the realm of the silver screen and television. In films and TV shows, characters often find themselves gazing upward during meteor showers, a cinematic portrayal of humanity's universal connection to the cosmos. From science fiction epics to heartwarming dramas, the Perseids' radiant presence adds a touch of magic to on-screen narratives.

In contemporary society, the Perseid meteor shower has become more than just a celestial event—it's a cultural touchstone that brings people together. Festivals, gatherings, and public viewing events celebrate the Perseids' brilliance, fostering a sense of community and shared wonder. Social media platforms have amplified the reach of the Perseids, enabling individuals from around the world to share their experiences, photographs, and reflections on the cosmic display.

The Perseid meteor showers' journey through popular culture is a testament to humanity's ongoing dialogue with the cosmos. The shooting stars that grace our skies serve as a constant source of inspiration, sparking creativity, sparking curiosity, and kindling a sense of unity among individuals from all walks of life enabling individuals from around the world to share their experiences, photographs, and reflections on the cosmic display.

The Perseid meteor showers' journey through popular culture is a testament to humanity's ongoing dialogue with the cosmos. The shooting stars that grace our skies serve as a constant source of inspiration, sparking creativity, sparking curiosity, and kindling a sense of unity among individuals from all walks of life.

Perseids and Space Exploration

Beyond its role as a visual spectacle in our night sky, the Perseid meteor shower has also played an unexpected yet significant part in advancing our understanding of space.

Meteor showers like the Perseids provide scientists with a unique opportunity to study cosmic materials up close. As meteoroids enter our atmosphere and burn up, they release a spectrum of light that can be analyzed to determine their composition. By studying the chemical makeup of meteoroids, researchers gain insights into the composition of comets, asteroids, and other celestial bodies. This invaluable information informs our understanding of the early solar system and the building blocks of the universe.

Comet Swift-Tuttle, the progenitor of the Perseid meteor shower, carries within it a treasure trove of information about the solar system's origins. Studying the composition of Swift-Tuttle's nucleus and the meteoroids it sheds light on the conditions that prevailed during the formation of our cosmic neighborhood. By unraveling the comet's secrets, scientists gain clues about the materials that formed planets, moons, and other celestial objects.

Meteor showers have left their imprint on space missions, both planned and unplanned. Spacecraft entering Earth's atmosphere during meteor showers can be temporarily illuminated by the meteors, creating brilliant streaks visible from the ground. This phenomenon, known as "shooting stars from space," captivates observers and adds an unexpected cosmic element to spaceflight. Additionally, meteoroid impacts on spacecraft provide valuable data about the space environment and the potential hazards faced by vehicles traveling beyond our planet.

Meteor showers have practical implications for space exploration beyond their scientific value. By studying meteoroids and their interaction with Earth's atmosphere, scientists gain insights into the challenges posed by space debris—man-made remnants from past space missions. Understanding how meteoroids burn up and disintegrate helps inform strategies for managing and mitigating the risks posed by orbital debris, safeguarding future space exploration endeavors

The Perseid meteor shower's influence on space exploration extends beyond its scientific contributions. Just as the Perseids have inspired generations of stargazers, they also fuel the dreams of aspiring astronauts, engineers, and scientists.

The shooting stars that streak across the night sky serve as a reminder of the uncharted territory that awaits humanity in the cosmos, inspiring the next generation to reach for the stars and explore the mysteries of the universe.

Astronomy enthusiasts and professional scientists often collaborate to observe and document meteor showers, including the Perseids. Citizen science initiatives, such as crowd-sourced meteor counts, contribute valuable data to scientific research. Space agencies, such as NASA and ESA, study meteoroids to enhance our understanding of space dynamics and ensure the safety of future space missions.

As we journey through the realm of space exploration, the Perseid meteor shower emerges not only as a radiant display but also as a cosmic partner in our quest to comprehend the universe. Its meteoroids carry tales of the distant past, its brilliance illuminates the darkness of space, and its interaction with Earth's atmosphere offers insights into the intricacies of the cosmos.

Citizen Science and Perseids

In the collaborative symphony of scientific exploration, amateur astronomers and passionate skywatchers play a vital role.

Citizen science is a powerful force that brings together individuals from diverse backgrounds who share a common love for the cosmos. Amateur astronomers and skywatchers play a crucial role in observing and recording data during meteor showers like the Perseids. Their keen eyes, paired with modern technology, contribute to the wealth of information collected about the shower's intensity, meteor counts, and unique characteristics.

Citizen scientists harness their passion for the night sky to engage in crowd-sourced data collection during meteor showers. Through meteor counting efforts, enthusiasts worldwide contribute to databases that document the number, brightness, and trajectory of meteors during peak shower periods. These combined observations provide a broader perspective of the meteor shower's behavior, enabling scientists to analyze trends and fluctuations across different regions.

The collective observations of citizen scientists have enriched our understanding of meteor showers, shedding light on variables such as meteor rates, brightness, and even potential variations from year to year. These contributions are invaluable in refining scientific models that predict meteor shower behavior and allow researchers to more accurately forecast peak meteor activity. Citizen science efforts amplify the scientific impact of meteor shower observations, turning the night sky into an interactive laboratory.

Participating in citizen science initiatives empowers enthusiasts of all levels, transforming them into contributors to genuine scientific discovery. Amateurs with varying degrees of expertise can engage with meteor shower observations, from beginners identifying the brighter meteors to seasoned skywatchers recording comprehensive data sets. This inclusivity bridges the gap between amateur enthusiasts and professional scientists, fostering collaboration and mutual learning.

The Perseid meteor shower stands as a prime case study in the power of citizen science. Through various online platforms, meteor shower enthusiasts collaborate in real-time, sharing their observations and creating a global snapshot of the meteoroid stream's interaction with Earth's atmosphere. By analyzing this collective data, researchers gain insights into the spatial and temporal distribution of meteoroids, contributing to a deeper understanding of the shower's dynamics. Citizen science initiatives extend beyond scientific data collection—— they foster a deeper connection between individuals and the cosmos. Engaging in collaborative efforts to study meteor showers transforms observers into active participants, nurturing a sense of awe and wonder for the night sky. As citizen scientists gaze upward, they not only contribute to scientific knowledge but also forge a lasting bond with the universe.

In the realm of meteor shower observation, the boundary between amateur and professional blurs, as both groups join forces to unravel the mysteries of the cosmos. Citizen scientists serve as vital partners in expanding our knowledge of the Perseid meteor shower and beyond, showcasing the transformative power of collective curiosity and collaboration.

Exploring the Future of the Perseid Meteor Shower

As time marches forward, the Perseid meteor shower continues its eternal dance across the celestial stage.

The Perseid meteor showers' future behavior remains a cosmic puzzle waiting to be solved. As scientists refine their understanding of the meteoroid stream's dynamics, they seek to predict variations in meteor rates, brightness, and potential deviations from historical patterns. The quest to uncover the reasons behind these fluctuations fuels ongoing research, shedding light on the intricate dance between Earth, Swift-Tuttle, and the cosmos.

The Perseid meteor shower's behavior can be influenced by a variety of cosmic factors, from the gravitational perturbations of other celestial bodies to interactions with space debris and

other meteoroid streams. As Earth's orbit changes and Swift-Tuttle's path evolves, the timing and intensity of the Perseids may undergo shifts. The cosmic choreography that shapes the meteoroid stream's trajectory underscores the complex interplay between the forces that govern our universe. As humanity's footprint on Earth grows, the impact of artificial light pollution becomes increasingly relevant to meteor shower observations. Finding dark skies for optimal viewing becomes more challenging, necessitating efforts to preserve and protect areas where the night sky remains relatively untainted. Additionally, the presence of satellites and space debris in Earth's orbit introduces new challenges and opportunities for meteoroid interactions, potentially altering the Perseid meteor showers' characteristics.

In an era of rapid technological advancement, the Perseid meteor showers' allure remains timeless. With each passing year, more people are exposed to the beauty and wonder of the cosmos through digital media, fostering renewed cultural interest in astronomical events. This resurgence in curiosity may lead to increased participation in observing the Perseids and other celestial phenomena, creating a new generation of stargazers and skywatchers.

The rise of space tourism introduces a new dimension to meteor shower observation. As commercial space travel becomes more accessible, individuals may have the opportunity to witness meteor showers from an entirely different vantage point—above Earth's atmosphere. Viewing the Perseids from the vantage of space adds a layer of uniqueness to the experience, offering a fresh perspective on the cosmic spectacle that has captivated humanity for generations. While the future of the Perseid meteor shower remains shrouded in mystery, its enduring legacy is clear. From inspiring ancient myths to advancing scientific knowledge, the Perseids continue to ignite human imagination and curiosity. As we gaze toward the stars and await each annual meteor shower, we stand as witnesses to the ongoing cosmic dance—a dance that transcends time, space, and the boundaries of human understanding.

How to Photograph the Perseids

The Perseid meteor shower, a grand spectacle in the night sky, offers a captivating visual feast that extends beyond the naked eye.

Photographing the Perseids requires meticulous planning. Select a location far from urban light pollution to ensure optimal visibility of the meteor shower. Familiarize yourself with the night sky, identifying the radiant point in the Perseus constellation where the meteors seem to originate. Choose the right equipment, including a sturdy tripod to keep your camera steady during long exposures. A digital single-lens reflex (DSLR) or mirrorless camera with manual settings is ideal for capturing meteor showers. Use a wide angle lens with a fast aperture to capture as much of the night sky as possible and gather more light. Set your camera to manual mode, enabling you to adjust key settings such as aperture, shutter speed, and ISO. Experiment with higher ISO settings to enhance the sensitivity of your camera's sensor.

Long exposures are key to capturing meteor trails. Set your camera to Bulb mode or manual shutter speed and experiment with exposure times ranging from a few seconds to several minutes. Keep in mind that longer exposures increase the likelihood of capturing multiple meteors in a single frame. Be patient, as meteor activity can be sporadic, and waiting for the perfect shot is part of the process.

Compose your shots to include a visually interesting foreground— whether it's a landscape, silhouetted trees, or a tranquil lake. Utilize the rule of thirds to create a balanced composition, placing the radiant point or a recognizable constellation off-center. Frame the sky in a way that captures a generous portion of the celestial canvas, allowing for the meteors' unpredictable trajectories.

Once you've captured your shots, the post-processing stage brings out the full potential of your images. Use photo editing software to adjust exposure, contrast, and color balance. Experiment with enhancing the visibility of meteor trails by adjusting the brightness and contrast levels. Stack multiple images to create composite shots, emphasizing the meteor shower's magnificence.

As your meteor shower photographs come to life, share your cosmic creations with fellow sky enthusiasts and the broader community. Posting on social media platforms, astronomy forums, or personal blogs connects you with like-minded individuals who share your passion for the night sky.

Your images may inspire others to venture out and experience the Perseids for themselves, perpetuating the cycle of cosmic curiosity.

Photographing the Perseids transcends the technical aspects of photography; it's an endeavor that captures the magic and wonder of the cosmos. Each image becomes a snapshot of celestial storytelling, a testament to your connection with the universe. Whether you're a seasoned astrophotographer or embarking on your first meteor shower shoot, the process of capturing cosmic fireworks adds a dimension of awe and beauty to your visual journey.

Exploring Other Meteor Spectacles

While the Perseid meteor shower reigns as one of the most famous and anticipated celestial events, it is not alone in the night sky's cosmic theater.

Geminids: A December Delight

As the year draws to a close, the Geminid meteor shower paints the December sky with its luminous display. Originating from the asteroid 3200 Phaethon, this shower produces a striking number of bright, slow-moving meteors that captivate observers. The Geminids' origin from an asteroid rather than a comet sets them apart, igniting scientific curiosity about the connection between asteroids and meteor showers.

Quadrantids: A New Year's Celestial Celebration

In the frosty nights of early January, the Quadrantid meteor shower welcomes the new year with its brilliant streaks. The origin of the Quadrantids traces back to an extinct comet or an asteroid named 2003 EH1. Known for its brief but intense bursts of activity, the Quadrantids illuminate the night sky with its fast-moving meteors, providing a dazzling spectacle to start the year.

Leonids: Cosmic Showers from Comet Tempel-Tuttle

The Leonid meteor shower, named after the constellation Leo, is associated with the periodic comet Tempel-Tuttle. Known for its occasional meteor storms, the Leonids have produced some of the most remarkable displays in history, with meteors raining down in breathtaking numbers. These meteor storms have left observers in awe and sparked scientific investigations into the dynamic behavior of meteoroid streams.

Orionids: Halley's Cosmic Debris

The Orionid meteor shower, originating from Halley's Comet, graces the skies of October. These fast-moving meteors are known for their brilliance and distinctive trails, captivating stargazers as they streak across the night sky. The connection to Halley's Comet adds an element of continuity to the celestial tapestry, as the comet's debris continues to create cosmic fireworks long after its fleeting appearances.

Eta Aquarids: The Beauty of Halley's Farewell

In May, the Eta Aquarid meteor shower arrives as a farewell gift from Halley's Comet. These meteors light up the pre-dawn sky, leaving trails of luminescence that linger in the early morning darkness. The Eta Aquarids serves as a reminder of the cometary travelers that venture through our solar system, leaving behind cosmic breadcrumbs that dazzle as they burn up in Earth's atmosphere.

As we explore the myriad meteor showers that grace our night skies, we recognize the harmony of the cosmos—a symphony of celestial events that remind us of our connection to the universe. Each meteor shower tells a unique story, reflecting the journeys of comets, asteroids, and other cosmic wanderers. As we gaze upward, we become witnesses to these celestial tales, marveling at the grandeur and mystery of the universe.

CONCLUSION

The Celestial Ballet: Perseid Meteor Shower and our Cosmic Connection

As we reach the conclusion of our cosmic voyage through the pages of this book, we find ourselves standing at the crossroads of science, culture, and wonder. The Perseid meteor shower, a radiant gift from the universe, has been our guide, leading us through the intricate tapestry of its origins, its impact on humanity, and its enduring legacy in the night sky.

From the ancient myths that breathe life into constellations to the scientific revelations that unveil the mysteries of space, the Perseids have been our companions on a journey of discovery. We have ventured through time and space to uncover the dance of debris left by Comet Swift-Tuttle, a journey that echoes the rhythmic pulse of the cosmos.

With each meteor streaking across the sky, we witnessed more than just the fleeting beauty of a shooting star. We glimpsed our connection to the universe, a reminder that we are part of a grand cosmic narrative. The Perseids have inspired astronomers, artists, poets, and dreamers for generations, leaving their indelible mark on human culture and imagination. But our exploration does not end here.

As we conclude this book, we invite you to step outside on a clear night, gaze upward, and imagine the countless meteoroids journeying through space. The Perseid meteor shower continues its timeless performance, inviting us to be witnesses to its celestial ballet.

As you immerse yourself in the beauty of the night sky, remember that the Perseids are not just astronomical events; they are invitations to ponder the universe's mysteries and celebrate the interconnectedness of all things. Let the Perseids remind us that we are part of a vast cosmic story——one that unfolds with every shooting star and every gaze heavenward. May the Perseid meteor shower and its luminous trails continue to spark awe and curiosity in your heart, inspiring you to explore the universe both within and beyond. Our journey together through the Perseids may have come to an end, but the cosmic dance carries on, inviting us to look up, dream, and embrace the stars.

In the end, it's not just about observing meteor showers—it's about discovering the universe and discovering ourselves in the process. The universe is vast, mysterious, and ever-changing, and we are fortunate to be part of this grand cosmic ballet.

May your nights be clear, your eyes be open to the wonders above, and your heart forever connected to the stars.

The end is not the end; it's a new beginning—a beginning filled with stardust and infinite possibilities. Thank you for joining us on this journey.

Keep looking up.

EXTRAS

Perseid Meteor Shower Calendar

- **Mid-July to Late August:** The Perseid meteor shower is active during this period, with the peak occurring around mid-August.
- **July 17 - August 24, 2023:** In 2023, the Perseids will grace the night sky, offering stargazers the chance to witness their spectacular display.
- **Peak Nights:** The Perseids are expected to be most active between the night of August 12 and the early hours of August 13. Mark this on your calendar for optimal viewing!

EXTRAS

Glossary of Terms

- **Meteoroid:** A small rocky or metallic body in outer space, often a fragment of a larger celestial body.
- **Meteor:** A bright streak of light produced by a meteoroid entering Earth's atmosphere and burning up due to friction.
- **Meteor Shower:** A phenomenon where a large number of meteors appear to originate from a single point in the sky, caused by Earth passing through the debris left behind by a comet or asteroid.
- **Radiant:** The point in the sky from which meteors in a meteor shower appear to originate.
- **Zenithal Hourly Rate (ZHR):** The number of meteors an observer would see in an hour under ideal conditions, with the radiant directly overhead (zenith).

EXTRAS

Resources for Further Reading

1. **American Meteor Society:** Visit the AMS website for comprehensive information about meteor showers, including observing tips, calendars, and educational resources. (Website: www.amsmeteors.org)
2. **International Meteor Organization (IMO):** The IMO offers global information on meteor showers, meteor observing techniques, and international collaboration. (Website: www.imo.net)
3. **NASA's Meteoroid Environment Office:** Explore NASA's resources on meteoroids, meteor showers, and space weather. (Website:www.nasa.gov/content/meteoroid-environment-office)
4. **Sky & Telescope:** A reputable source for skywatching and astronomy enthusiasts, offering guides to meteor showers and celestial events. (Website:www.skyandtelescope.org)
5. **Stellarium:** Use this planetarium software to simulate the night sky and plan your meteor shower observations. Available as a desktop program and mobile app. (Website:stellarium.org)
6. **Astronomy Clubs and Societies:** Connect with local astronomy clubs and societies to stay updated on upcoming meteor showers and participate in group observations.
7. **Books on Meteor Showers:** Explore books like *"Meteor Showers and their Parent Comets"* by Peter Jenniskens and *"The Practical Astronomer's Deep-sky Companion"* by Jess K. Gilmour for in-depthknowledge.
8. **Online Communities:** Engage with fellow astronomy enthusiasts on online forums, subreddits, and social media platforms to share experiences and tips for observing meteor showers.

STARDUST SERENADE

www.ingramcontent.com/pod-product-compliance
Lightning Source LLC
Chambersburg PA
CBHW072225290526
45794CB00007B/2891